Laboratory Manual
HUMAN BIOLOGY
AND HEALTH

Prentice Hall
Englewood Cliffs, New Jersey
Needham, Massachusetts

Laboratory Manual

PRENTICE HALL SCIENCE
Human Biology and Health

ISBN 0-13-986795-3

3 4 5 6 7 8 9 10   96 95

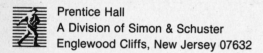
Prentice Hall
A Division of Simon & Schuster
Englewood Cliffs, New Jersey 07632

Contents

Safety Symbols   H5

Science Safety Rules   H6

CHAPTER 1 ■ *The Human Body*

1   Investigating Tissues   H9

CHAPTER 2 ■ *Skeletal and Muscular Systems*

2   Examining Bones and Joints   H13

3   Comparing Bones, Joints, and Muscles   H17

CHAPTER 3 ■ *Digestive System*

4   Nutrient Identification   H23

5   Observing the Digestion of Starches   H29

6   Investigating Proteins   H33

7   How Much Vitamin C Is in Fruit Juice?   H39

CHAPTER 4 ■ *Circulatory System*

8   Observing Blood Circulation   H43

9   Investigating the Heart   H47

CHAPTER 5 ■ *Respiratory and Excretory Systems*

10   Investigating Breathing and Respiration   H51

11   Investigating the Effect of Exercise
on Respiration   H57

CHAPTER 6 ■ *Nervous and Endocrine Systems*

12   Observing Human Reflexes   H61
13   Observing the Effect of Adrenaline on *Daphnia*   H65

CHAPTER 7 ■ *Reproduction and Development*

14   Observing the Development of Frog Eggs   H69

CHAPTER 8 ■ *Immune System*

15   A Model for Disease Transmission   H73
16   Relating Noninfectious Diseases and Nutrition   H77

CHAPTER 9 ■ *Alcohol, Tobacco, and Drugs*

17   Effects of Tobacco and Alcohol on Seed
     Germination   H85

Safety Symbols

All the investigations in this *Laboratory Manual* have been designed with safety in mind. If you follow the instructions, you should have a safe and interesting year in the laboratory. Before beginning any investigation, make sure you read the safety rules that follow.

The eight safety symbols below appear next to certain steps in some of the investigations in this *Laboratory Manual*. The symbols alert you to the need for special safety precautions. The description of each symbol below tells you which precautions to take whenever you see the symbol in an investigation.

 Glassware Safety

1. Whenever you see this symbol, you will know that you are working with glassware that can easily be broken. Take particular care to handle such glassware safely. And never use broken or chipped glassware.
2. Never heat glassware that is not thoroughly dry. Never pick up any glassware unless you are sure it is not hot. If it is hot, use heat-resistant gloves.
3. Always clean glassware thoroughly before putting it away.

 Fire Safety

1. Whenever you see this symbol, you will know that you are working with fire. Never use any source of fire without wearing safety goggles.
2. Never heat anything—particularly chemicals—unless instructed to do so.
3. Never heat anything in a closed container.
4. Never reach across a flame.
5. Always use a clamp, tongs, or heat-resistant gloves to handle hot objects.
6. Always maintain a clean work area, particularly when using a flame.

 Heat Safety

Whenever you see this symbol, you will know that you should put on heat-resistant gloves to avoid burning your hands.

 Chemical Safety

1. Whenever you see this symbol, you will know that you are working with chemicals that could be hazardous.
2. Never smell any chemical directly from its container. Always use your hand to waft some of the odors from the top of the container toward your nose—and only when instructed to do so.
3. Never mix chemicals unless instructed to do so.
4. Never touch or taste any chemical unless instructed to do so.
5. Keep all lids closed when chemicals are not in use. Dispose of all chemicals as instructed by your teacher.
6. Immediately rinse with water any chemicals, particularly acids, that get on your skin and clothes. Then notify your teacher.

 Eye and Face Safety

1. Whenever you see this symbol, you will know that you are performing an experiment in which you must take precautions to protect your eyes and face by wearing safety goggles.
2. When you are heating a test tube or bottle, always point it away from you and others. Chemicals can splash or boil out of a heated test tube.

 Sharp Instrument Safety

1. Whenever you see this symbol, you will know that you are working with a sharp instrument.
2. Always use single-edged razors; double-edged razors are too dangerous.
3. Handle any sharp instrument with extreme care. Never cut any material toward you; always cut away from you.
4. Immediately notify your teacher if your skin is cut.

 Electrical Safety

1. Whenever you see this symbol, you will know that you are using electricity in the laboratory.
2. Never use long extension cords to plug in any electrical device. Do not plug too many appliances into one socket or you may overload the socket and cause a fire.
3. Never touch an electrical appliance or outlet with wet hands.

 Animal Safety

1. Whenever you see this symbol, you will know that you are working with live animals.
2. Do not cause pain, discomfort, or injury to an animal.
3. Follow your teacher's directions when handling animals. Wash your hands thoroughly after handling animals or their cages.

Science Safety Rules

One of the first things a scientist learns is that working in the laboratory can be an exciting experience. But the laboratory can also be quite dangerous if proper safety rules are not followed at all times. To prepare yourself for a safe year in the laboratory, read over the following safety rules. Then read them a second time. Make sure you understand each rule. If you do not, ask your teacher to explain any rules you are unsure of.

Dress Code

1. Many materials in the laboratory can cause eye injury. To protect yourself from possible injury, wear safety goggles whenever you are working with chemicals, burners, or any substance that might get into your eyes. Never wear contact lenses in the laboratory.
2. Wear a laboratory apron or coat whenever you are working with chemicals or heated substances.
3. Tie back long hair to keep it away from any chemicals, burners, and candles, or other laboratory equipment.
4. Remove or tie back any article of clothing or jewelry that can hang down and touch chemicals and flames.

General Safety Rules

5. Read all directions for an experiment several times. Follow the directions exactly as they are written. If you are in doubt about any part of the experiment, ask your teacher for assistance.
6. Never perform activities that are not authorized by your teacher. Obtain permission before "experimenting" on your own.
7. Never handle any equipment unless you have specific permission.
8. Take extreme care not to spill any material in the laboratory. If a spill occurs, immediately ask your teacher about the proper cleanup procedure. Never simply pour chemicals or other substances into the sink or trash container.
9. Never eat in the laboratory.
10. Wash your hands before and after each experiment.

First Aid

11. Immediately report all accidents, no matter how minor, to your teacher.
12. Learn what to do in case of specific accidents, such as getting acid in your eyes or on your skin. (Rinse acids from your body with lots of water.)
13. Become aware of the location of the first-aid kit. But your teacher should administer any required first aid due to injury. Or your teacher may send you to the school nurse or call a physician.
14. Know where and how to report an accident or fire. Find out the location of the fire extinguisher, phone, and fire alarm. Keep a list of important phone numbers—such as the fire department and the school nurse —near the phone. Immediately report any fires to your teacher.

Heating and Fire Safety

15. Again, never use a heat source, such as a candle or a burner, without wearing safety goggles.
16. Never heat a chemical you are not instructed to heat. A chemical that is harmless when cool may be dangerous when heated.
17. Maintain a clean work area and keep all materials away from flames.
18. Never reach across a flame.
19. Make sure you know how to light a Bunsen burner. (Your teacher will demonstrate the proper procedure for lighting a burner.) If the flame leaps out of a burner toward you, immediately turn off the gas. Do not touch the burner. It may be hot. And never leave a lighted burner unattended!
20. When heating a test tube or bottle, always point it away from you and others. Chemicals can splash or boil out of a heated test tube.

21. Never heat a liquid in a closed container. The expanding gases produced may blow the container apart, injuring you or others.
22. Before picking up a container that has been heated, first hold the back of your hand near it. If you can feel the heat on the back of your hand, the container may be too hot to handle. Use a clamp or tongs when handling hot containers.

Using Chemicals Safely

23. Never mix chemicals for the "fun of it." You might produce a dangerous, possibly explosive substance.
24. Never touch, taste, or smell a chemical unless you are instructed by your teacher to do so. Many chemicals are poisonous. If you are instructed to note the fumes in an experiment, gently wave your hand over the opening of a container and direct the fumes toward your nose. Do not inhale the fumes directly from the container.
25. Use only those chemicals needed in the activity. Keep all lids closed when a chemical is not being used. Notify your teacher whenever chemicals are spilled.
26. Dispose of all chemicals as instructed by your teacher. To avoid contamination, never return chemicals to their original containers.
27. Be extra careful when working with acids or bases. Pour such chemicals over the sink, not over your workbench.
28. When diluting an acid, pour the acid into water. Never pour water into the acid.
29. Immediately rinse with water any acids that get on your skin or clothing. Then notify your teacher of any acid spill.

Using Glassware Safely

30. Never force glass tubing into a rubber stopper. A turning motion and lubricant will be helpful when inserting glass tubing into rubber stoppers or rubber tubing. Your teacher will demonstrate the proper way to insert glass tubing.
31. Never heat glassware that is not thoroughly dry. Use a wire screen to protect glassware from any flame.
32. Keep in mind that hot glassware will not appear hot. Never pick up glassware without first checking to see if it is hot. See #22.
33. If you are instructed to cut glass tubing, fire-polish the ends immediately to remove sharp edges.
34. Never use broken or chipped glassware. If glassware breaks, notify your teacher and dispose of the glassware in the proper trash container.
35. Never eat or drink from laboratory glassware.
36. Thoroughly clean glassware before putting it away.

Using Sharp Instruments

37. Handle scalpels or razor blades with extreme care. Never cut material toward you; cut away from you.
38. Immediately notify your teacher if you cut your skin when working in the laboratory.

Animal Safety

39. No experiments that cause pain, discomfort, or harm to mammals, birds, reptiles, fish, and amphibians should be done in the classroom or at home.
40. Animals should be handled only if necessary. If an animal is excited or frightened, pregnant, feeding, or with its young, special handling is required.
41. Your teacher will instruct you as to how to handle each animal species that may be brought into the classroom.
42. Clean your hands thoroughly after handling animals or the cage containing animals.

End-of-Experiment Rules

43. After an experiment has been completed, clean up your work area and return all equipment to its proper place.
44. Wash your hands after every experiment.
45. Turn off all candles and burners before leaving the laboratory. Check that the gas line leading to the burner is off as well.

_____ *Laboratory Investigation* _____

Investigating Tissues

Background Information

There are four types of tissues in the human body: muscle, connective, nerve, and epithelial. Muscle tissue moves body parts. Connective tissue supports the body and unites some of its parts. Nerve tissue carries messages back and forth between the brain and spinal cord and every part of the body. Epithelial tissue forms a protective surface on the outside of the body and lines many cavities within the body.

In this investigation you will observe the four types of tissues in a chicken wing.

Problem

What are the four types of tissues?

Materials *(per group)*

chicken wing
paper towels
scissors
dissecting tray
dissecting needle

Procedure

1. Obtain the chicken wing from your teacher.

2. Rinse the chicken wing under running water and thoroughly dry it with paper towels. Place the chicken wing in a dissecting tray.

3. Examine the skin covering the chicken wing.

4. Remove the skin from the wing using the scissors. **CAUTION:** *Be careful when using scissors.* Carefully cut the skin along the entire length of the chicken wing as shown in Figure 1. Try not to cut through the muscles located below the skin.

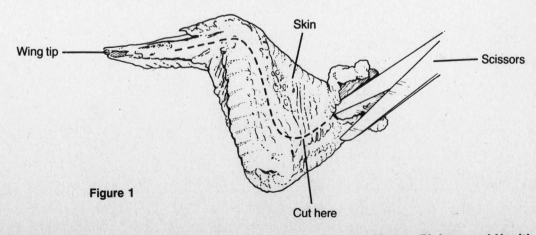

Figure 1

5. Notice the yellowish tissue found in small clumps on the inside of the skin. This tissue is a type of connective tissue called fat.

6. Observe the muscles on the chicken wing. The muscles are bundles of pale pink tissue that surround the bone.

7. Observe the shiny white tissue, or tendons, at the ends of the muscles. Tendons attach muscle to bone.

8. Notice the whitish tissue, or ligaments, between the bones. Ligaments hold bones together.

9. Locate a thin, white strand of material with the dissecting needle. Carefully pull the strand aside with the dissecting needle. This strand is a nerve.

10. Notice a thin reddish-brown strand of tissue. Pull it aside with the dissecting needle. This strand is a blood vessel.

11. In Figure 2, label a tendon, a muscle, and a bone.

12. Thoroughly wash your hands with soap and water.

13. Return all equipment to the storage area. Return the chicken wing to your teacher.

Observations

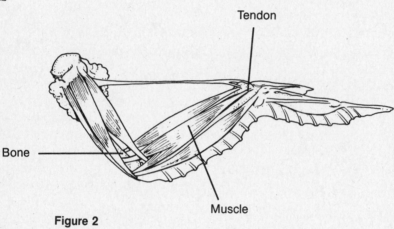

Tendon

Bone

Muscle

Figure 2

Analysis and Conclusions

1. Identify the following types of tissues as connective, muscle, nerve, or epithelial.

a. Tendon _____

b. Nerve _____

c. Fat _____

d. Blood vessel _____

e. Skin _____

f. Ligament _____

g. Bone _____

h. Muscle _____

2. Why are tendons important to a muscle's ability to move a bone?

_____

_____

_____

3. What enables the chicken to move its wing? _____

_____

Critical Thinking and Application

1. Which tissue of the chicken wing is commonly referred to as the meat?

_____

2. Why would a bird be unable to fly if there was some damage to the nerve in the wing?

_____

_____

_____

Going Further

Obtain prepared slides of each of the four types of tissues. Place them under the microscope and observe under the low- and high-power objectives. Draw what you see. Compare each type of tissue.

_____ *Laboratory Investigation* _____

Skeletal and Muscular Systems _____ **2** _____

Examining Bones and Joints

Background Information

The human skeleton is made up of 206 bones. In addition to giving shape, support, and protection to the body, bones store minerals such as calcium and phosphorus. These minerals make bones hard. Bone is also made up of living material. The tough membrane surrounding the long bone helps to supply the bone's living tissue with food and oxygen. In the center of bone is a soft red or yellow material called marrow. Red marrow produces blood cells. Yellow marrow contains fat and blood vessels.

Together with the action of the muscular system, bones make movement possible. Movement occurs when muscles, which are attached by tendons to bones, contract and relax. The place where two bones meet is called a joint. There are three basic types of joints: hinge, pivot, and ball-and-socket. Joints allow for movement of bones. Bones are attached to other bones by a connective tissue called a ligament.

In this investigation you will examine the structures of a long bone and determine how joints provide movement of bones.

Problem

What are the structures of a long bone? How do joints provide movement of bones?

Materials *(per group)*

dissecting needle
long bone from chicken leg
hand lens
scalpel
movable joint

Procedure
Part A   Examining a Long Bone

1. Obtain the long bone from your teacher.

2. Locate the smooth outer covering at the ends of the bone. Using the hand lens, notice the presence of small holes along the surface of the bone.

3. Using the dissecting needle, gently remove a portion of the membrane that surrounds the bone. **CAUTION:** *Be careful when using a dissecting needle.*

4. With a scalpel, cut one end of the bone lengthwise so you can see some of the red marrow. **CAUTION:** *Be careful when using a scalpel.*

5. Locate the compact and spongy areas of the bone.

6. In Figure 1, label the membrane surrounding the bone, marrow, compact bone, and spongy bone.

Part B   Examining the Movement of Joints

1. Obtain a movable joint from your teacher.

2. Carefully move the bones. Observe the type of movement.

3. Draw and identify the type of joint in the appropriate place in Observations.

Observations

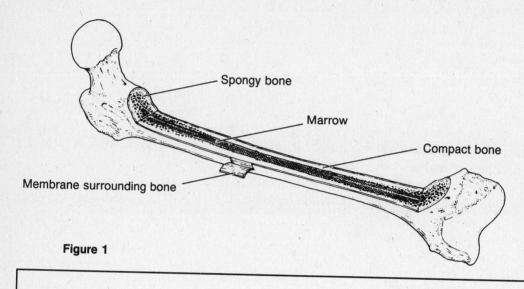

Spongy bone

Marrow

Compact bone

Membrane surrounding bone

Figure 1

Type of joint _____

Analysis and Conclusions

1. Describe each of the following parts of bone.

   a. Yellow marrow _____

   b. Spongy bone _____

   c. Compact bone _____

   d. Red marrow _____

2. What is the function of the small holes in the bone surface?

   _____

3. Explain why it is necessary for the ends of the bones to be smooth.

   _____

4. Why is bone considered living material? _____

   _____

Critical Thinking and Application

1. Why are there so many small bones in the hands and feet?

   _____

2. The joints of the skull are immovable. Why is this so?

   _____

   _____

   _____

3. a. Which type of joint—ball-and-socket, hinge, or pivot—allows the greatest range of

   movement? _____

   _____

   _____

   _____

   b. Where in the body is this type of joint found? _____

   _____

4. Suppose a person's diet lacks the mineral calcium. How would this deficiency affect the

skeletal system? _____

_____

_____

_____

Going Further

Examine a prepared slide of compact bone under the low-power objective lens of the microscope. Identify the canals and bone cells.

_____ *Laboratory Investigation* _____

Comparing Bones, Joints, and Muscles

Background Information

Bone is connective tissue containing both living and nonliving materials. Bone tissue may be either compact or spongy. Compact bone tissue is very dense and hard. Spongy bone tissue is softer and contains many spaces. The living part of bone tissue consists of bone cells surrounded by hard, nonliving materials such as the minerals calcium and phosphorus. Throughout compact bone tissue, there is a system of small canals. These canals contain blood vessels and nerves.

Together with the muscular system, bones make movement possible. Movement occurs when the muscles attached to bones contract and relax. The bones are connected at junctions called joints.

The human body has three types of muscle tissue. Skeletal muscle, or striated muscle, is used to move such body parts as the arms and legs. Skeletal muscles are also called voluntary muscles, because they can be moved at will. The second type of muscle, smooth muscle, is found in the walls of the digestive tract and of some blood vessels. The third type of muscle tissue, cardiac muscle, is found only in the heart. Cardiac and smooth muscles are involuntary and contract automatically.

In this investigation you will study the internal structure of the bone, compare the movements of various joints, and compare the three types of muscle tissue.

Problem

What is the internal structure of bone? How do various joints move? What are the similarities and differences among the three types of muscle tissue?

Materials *(per group)*

microscope
prepared slides of human
   compact bone
prepared slides of human skeletal,
   smooth, and cardiac muscle

Procedure
Part A Bone Tissue

🔺 1. Observe a prepared slide of a compact bone section under the low power of the microscope.

2. Switch to high power, and locate the canals. Observe how layers of bone encircle the canal. Locate bone cells, within the spaces between the layers of bone. See Figure 1. Draw and label what you see in the appropriate space in Observations.

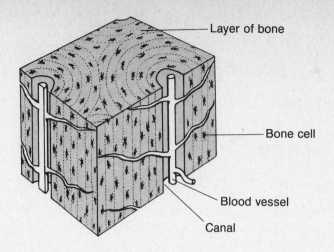

Figure 1

Part B Joints and Movements

1. Study Figure 2, which shows the various movements that joints permit.

2. Have your partner try to move each of the joints listed in the Data Table. For each joint, place a plus sign (+) in the appropriate box to indicate that a type of movement is possible. Use a minus sign (−) to indicate that a movement is not possible.

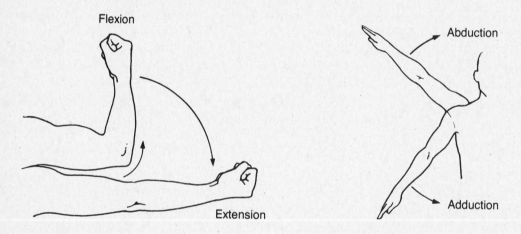

Flexion—the angle between the two bones decreases

Extension—the angle between the two bones increases

Abduction—a bone is moved away from the midline of the body

Adduction—a bone is moved toward the midline of the body

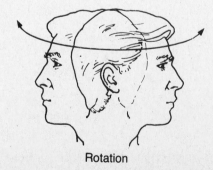

Rotation

Rotation—a body part partially revolves on its axis

Figure 2

Part C Muscle Tissue

1. Obtain prepared slides of skeletal, smooth, and cardiac muscle. Observe each slide under the low power of the microscope.

2. Switch to high power, and make a sketch of each muscle type in the appropriate space in Observations. For skeletal and cardiac muscle, label the nuclei, cytoplasm, and striations. For smooth muscle, label the nuclei, cytoplasm, and cell membranes.

Observations
Part A

PLATE 1

Magnification _____

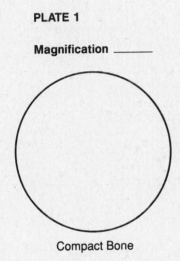

Compact Bone

Part B

DATA TABLE

Joint	Flexion	Extension	Adduction	Abduction	Rotation
Neck					
Shoulder					
Elbow					
Wrist					
Finger					
Hip					
Knee					
Ankle					

Part C

PLATE 2

Magnification _____

PLATE 3

Magnification _____

PLATE 4

Magnification _____

Smooth muscle

Skeletal muscle

Cardiac muscle

Analysis and Conclusions

1. a. What makes up the living tissue of bone? _____

_____

   b. What makes up the nonliving part? _____

_____

2. What is the function of the canals in bone tissue? _____

_____

_____

3. Which joint(s) permit the most possible movements? _____

_____

   The fewest? _____

4. Which of the various movements do all of the joints permit? _____

_____

5. a. How are skeletal muscle tissue and cardiac muscle tissue similar?

_____

   b. How are they different? _____

_____

_____

6. How is smooth muscle tissue different from the other two types of muscle tissue?

_____

_____

Critical Thinking and Application

1. Throughout a person's life, bone tissue is continually laid down and removed. However, as a person ages, the depositing of bony material slows. Use this information to explain the following facts:

   a. Elderly people often break bones when they fall. _____

   _____

   _____

   b. Broken bones heal more slowly in older people than in younger people.

   _____

   _____

2. Among athletes, injuries to the knee joint are fairly common. Suggest a reason for this

   based on what you learned in the investigation. _____

   _____

   _____

3. In what part of the body can each of the three types of muscle tissue be found?

   Skeletal muscle tissue _____

   Smooth muscle tissue _____

   Cardiac muscle tissue _____

Going Further

Using reference materials, find out the different types of joints and where they are found in the body. Indicate the type of joint used in the hip, knee, wrist, ankle, shoulder, elbow, and neck. Record this information in a data table.

_____ *Laboratory Investigation* _____

Nutrient Identification

Background Information

Among nutrients in food are carbohydrates (starches and sugars), fats, proteins, minerals, water, and vitamins. You can detect the presence of some of these nutrients by taste. For example, all foods that taste sweet contain some form of sugar unless they are artificially sweetened. On the other hand, some foods, such as milk and onions, contain sugar but do not taste sweet. Therefore, scientists do not rely only on taste or appearance to determine what nutrients a food contains. There are tests that can be used to aid in the identification of nutrients.

In this investigation you will learn to perform tests for the presence of starches, sugars, proteins, and fats.

Problem

What nutrients are found in various kinds of food?

Materials *(per group)*

samples of various foods
    including flour, honey, gelatin,
    and cooking oil
iodine solution
Benedict's solution
biuret solution
safety goggles
3 medicine droppers (1 for each
    solution)
2 test tubes
test-tube holder
hot-water bath (hot plate and
    beaker of water)
paper towels
brown wrapping paper or
    grocery bags

Procedure

Part A   Test for Starches

1. Place a small quantity of flour on a paper towel.

2. Using a medicine dropper, place 1 or 2 drops of iodine solution on the flour.
   CAUTION: *Keep iodine solution off your skin because it will leave a stain.*

3. Notice that the iodine solution turns a purplish-blue or blue-black color. This indicates that flour contains starch. If the iodine remains a yellow-brown color, starch is not present.

4. Test two to five other foods for starch and record your results in Data Table 1.

Part B Test for Sugars

1. Using a medicine dropper, place 30 drops of honey-and-water solution in a test tube.

2. Using another medicine dropper, add Benedict's solution until the test tube is about one third full. **CAUTION:** *Keep Benedict's solution away from your skin because it can burn you. If you spill some, rinse it off immediately with cold running water and inform your teacher.*

3. Place the test tube in the water bath and gently boil the mixture for 2 to 5 minutes. **CAUTION:** *Be careful when using the water bath. Don't allow the water to boil too vigorously. Be certain that the openings of the test tubes are not pointed toward people. Remember to wear your safety goggles.*

4. Using a test-tube holder, remove the test tube from the water bath. The solution should have turned green, yellow, orange, or orange-red. This indicates a positive test: sugar is present. If the Benedict's solution remains blue, the test is negative; sugar is not present. **Note:** *Benedict's solution indicates the presence of simple sugars such as glucose and fructose, which are found in most fruits.* It does not detect the presence of complex sugars such as lactose (milk sugar) and sucrose (table sugar).

5. Test two to five other foods for sugar and record your results in Data Table 2. If you use solid foods, crush the material to be tested, place it in a test tube, and add 30 drops of water.

Part C Test for Proteins

1. Using a medicine dropper, fill a test tube about one third full of gelatin solution.

2. Add 10 drops of biuret solution. **CAUTION:** *Biuret solution will burn skin and clothing.*

3. Hold the tube against a white background. Notice that the mixture has turned a violet color. This indicates the presence of protein. If there is no color change, protein is not present.

4. Test two to five other foods for protein and record your results in Data Table 3.

Part D Test for Fats

1. Place a few drops of cooking oil on a piece of brown paper bag. Place a few drops of water on another section of the same paper.

2. Hold the paper up to the light and look through the water spot and the fat spot. Note that the oil spot is greasy and translucent. This indicates the presence of fat.

3. Reexamine the spots after several minutes. The water spot will evaporate and disappear. The fat spot does not disappear.

4. Test two to five other foods for the presence of fat. If the food is a solid, rub it into the brown paper. Record your results in Data Table 4.

Observations

1. Record the results of your test for starches in the following table.

DATA TABLE 1

Food Tested	Color With Iodine Solution	Is Starch Present?

2. Record the results of your test for simple sugars in the following table.

DATA TABLE 2

Food Tested	Color With Benedict's Solution	Is Simple Sugar Present?

3. Record the results of your test for proteins in the following table.

DATA TABLE 3

Food Tested	Color With Biuret Solution	Is Protein Present?

4. Record the results of your test for fats in the following table.

DATA TABLE 4

Food Tested	Translucent Spot?	Is Fat Present?

Analysis and Conclusions

1. Of the foods you tested, which contain starch? How do you know?

_____

_____

_____

2. Of the foods you tested, which contain sugar? How do you know?

_____

_____

_____

3. Of the foods you tested, which contain protein? How do you know?

_____

_____

_____

4. Of the foods you tested, which contain fat? How do you know?

_____

_____

_____

Critical Thinking and Application

1. Why is it important that starches, sugars, proteins, and fats be included in a person's diet?

_____

_____

_____

2. If a food does not cause biuret solution to turn violet, do you know what type of food it

is? Explain your answer. _____

_____

_____

_____

Going Further

There are also chemical tests that detect the presence of different vitamins. Indophenol is a chemical that can be used to test for vitamin C. To conduct this test, pour indophenol into a test tube to a depth of 2 cm. Add the substance to be tested, one drop at a time. Keep track of the number of drops added and shake the test tube after each drop is added. Continue until the blue color disappears. The more drops of test substance required to bleach the indophenol, the less vitamin C the substance contains.

a. Compare the vitamin C content of various fruit juices, such as orange, apple, or grapefruit, or lemon, or various brands of one kind of juice.

b. Does cooking affect vitamin C content? To find out, test a substance before and after it is boiled.

_____ *Laboratory Investigation* _____

5

Observing the Digestion of Starches

Background Information

Digestion of starches begins in the mouth. The mouth contains salivary glands that produce about 1.5 L of saliva daily. Saliva is a mixture of water, mucus, and a tiny amount of a protein called ptyalin, which digests starch into sugar. Ptyalin is known as a digestive enzyme.

In this investigation you will learn what happens to starch in the mouth.

Problem

What does saliva do to starch in the mouth?

Materials *(per group)*

test-tube rack
3 test tubes, 10 cm × 1.5 cm,
    with test-tube stoppers
Benedict's solution
tripod
Bunsen burner
50-mL beaker
wire gauze
iodine solution
1% soluble starch solution
glass-marking pencil
10-mL graduated cylinder
safety goggles
paraffin wax

Procedure

1. Using the graduated cylinder, pour 5 mL of the starch solution into each of two test tubes. Label these test tubes 1 and 2. Then rinse the graduated cylinder with water.

2. Put three drops of iodine solution into each test tube. Stopper the test tubes. Gently shake the test tubes to mix the two solutions. The blue-black color of the mixture indicates the presence of starch. **CAUTION:** *Iodine solution will stain skin and clothing. Handle it carefully.*

3. Obtain a piece of paraffin wax. Have one person in the group rinse out his or her mouth with water and then chew the wax to increase the secretion of saliva. Collect about 2 mL of saliva in the graduated cylinder. Into test tube 1 only, add the 2 mL of saliva and gently shake the test tube to mix its contents.

4. Label a third test tube 3. Pour about half of the contents of test tube 1 into test tube 3. Add an equal amount of Benedict's solution to test tube 3. Cover the test tube with a stopper. Gently shake the test tube to mix the solution. **CAUTION:** *Keep the Benedict's solution away from your skin because it is caustic and can burn you. If you spill some, wash it off with cold water and inform your teacher.*

5. Remove the stopper. Heat test tube 3 in a boiling water bath for a few minutes. See Figure 1. **CAUTION:** *Wear safety goggles when heating substances, and be careful when using an open flame. Do not allow the contents of test tube 3 to come to a boil.* If simple sugars are present, the mixture will change to green, orange, or red after heating. Observe any changes that occur after heating.

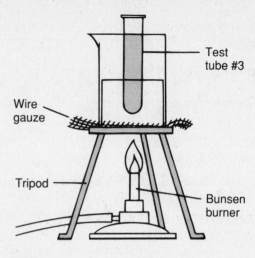

Figure 1

Observations

1. Describe what happens to the white starch solution when iodine solution is added to it.

_____

_____

2. What happened to the blue color of the Benedict's solution after test tube 3 was heated?

_____

_____

Analysis and Conclusions

1. What did the starch change to? _____

2. What happens to starch when it is mixed with saliva?

_____

3. What substance in saliva is responsible for the change in starch?

_____

4. What did you prove by putting the iodine solution into test tubes 1 and 2?

_____

_____

Critical Thinking and Application

1. What advantage is the action of saliva to your body?

_____

_____

_____

2. In addition to your mouth, where else in your body would you expect to find digestive

enzymes? _____

_____

_____

3. Design an experiment to determine if a change in temperature would affect the rate at

which a digestive enzyme works. _____

_____

_____

_____

_____

Going Further

Using reference materials, find out what techniques scientist use to remove digestive
juices from the organs of some animals. Experimentally, how would you test whether these
juices retained their strength?

_____ *Laboratory Investigation* _____

Investigating Proteins

Background Information

Proteins are part of all living things and are essential to life processes. All proteins are made of long chains of amino acids. Proteins contain nitrogen, carbon, hydrogen, and oxygen. Small amounts of iron, phosphorus, sulfur, and copper are often found in proteins. High-protein foods include fish, eggs, cheese, cereals, nuts, and meats.

Under certain conditions, the proteins in food substances can be made to coagulate, or clump or stick together. In this investigation you will determine what are some of the ways by which proteins can be made to coagulate.

Problem

How can the coagulation of protein molecules in flour, egg white, and milk be observed?

Materials (*per group*)

25-mL graduated cylinder	250-mL beaker	100-mL beaker
battery jar of water	2 watch glasses	skim milk, 15 mL
square of 6 thicknesses of	ring stand	(nonmodified type)
cheesecloth, 30 cm × 30 cm	support ring	medicine dropper
string, 15 cm long	wire gauze	test tube
spoon	Bunsen burner	test-tube rack
all-purpose flour, 55 g	forceps	dilute HCl
evaporating dish	egg white	

Procedure

Part A Coagulation of Proteins in Flour

⚗ **1.** Place 50 g of flour in a large evaporating dish. Add 25 mL of cold water and stir until a ball of dough forms.

2. Dust hands with flour and knead the dough for 5 minutes. Observe its appearance.

3. Place the ball of dough in the cheesecloth and secure the four ends with string, making a small bag.

4. Place the bag of dough in the battery jar of cold water, squeezing the bag to remove the starch. Change the water when it gets cloudy.

5. Repeat step 4, continuing to remove starch, until the water remains clear. Wash the bag of dough under running water for 2 minutes.

6. Place the bag in a watch glass. Then open the bag and remove the contents. Examine the dough. Record your observations.

7. Gently try to pull the ball of dough apart. Record your observations.

Part B   Coagulation of Proteins in Egg White

🔺 **1.** Set up the apparatus as shown in Figure 1.

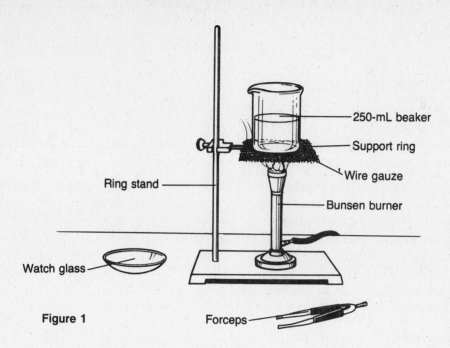

Figure 1

🔥 **2.** Fill the 250-mL beaker half-full with tap water. Heat the water until it boils.

🔥 **3.** As soon as the water boils, turn off the Bunsen burner and drop the egg white into the beaker of water. Observe for 2 minutes and record your observations.

4. Use the forceps to remove a piece of the egg white. Place it on the watch glass.

🔥 **5.** Reheat the egg white remaining in the beaker for 4 minutes. Turn off the burner at the end of this time and remove a piece of the egg white with the forceps. Place it on the watch glass next to the first sample. Compare the two samples. Record your observations.

Part C   Coagulation of Proteins in Milk

1. Pour 10 mL of skim milk into a test tube.

2. Add HCl drop by drop until a change is observable.

3. Let the test tube stand in the test-tube rack for 5 minutes.

4. Carefully decant (pour off) the liquid portion in the test tube into the beaker. Pour the solid portion onto the watch glass. Examine it and record your observations.

Observations
Part A

1. Describe the appearance of the dough as you knead it.

_____

_____

2. Describe the appearance of the contents of the bag in step 6.

_____

_____

3. What happens when you try to pull the ball of dough apart?

_____

_____

Part B

1. What is the appearance of the egg white during the first 2 minutes of heating?

_____

_____

_____

2. How do the two samples compare? _____

_____

_____

Part C

1. What happens as you add HCl to the milk? _____

_____

2. What is the appearance of the solid contents of the test tube?

_____

_____

3. Do the contents resemble any dairy food familiar to you? If so, what food?

_____

_____

Analysis and Conclusions
Part A

1. The material left in the bag after washing is called gluten.

   What makes up gluten? _____

2. Compare the original dough with gluten. _____

   _____

3. How can the protein in flour be made to coagulate?

   _____

   _____

Part B

How do you know that the proteins in the egg white have coagulated?

_____

_____

Part C

What happened to the milk when HCl was added? _____

_____

_____

Critical Thinking and Application

1. A substance is said to be soluble when it is able to be dissolved. Based on your

   observations, is starch soluble in water? Explain. _____

   _____

2. How would you explain the difference in texture between soft crumbly cakes and firm

   bread? _____

   _____

   _____

3. A poached egg is prepared by dropping an egg into boiling water. Explain why the water

   must be boiling. _____

   _____

   _____

4. In terms of coagulation of protein, explain why a soft-boiled egg needs less time to

prepare than a hard-boiled egg. _____

_____

5. Why does sour milk curdle? _____

_____

Going Further

Devise an experiment to determine if the addition of sugar, fat or oil, salt, and egg yolks increases or retards gluten formation in dough. Upon approval by your teacher, perform your experiment. Record your observations and conclusions.

_____ *Laboratory Investigation* _____

7 _____

Digestive System _____

How Much Vitamin C Is in Fruit Juice?

Background Information

Vitamin C is necessary to prevent scurvy and to help in the formation of teeth, bones, red blood cells, and connective tissue. The average teenager needs 100 mg of vitamin C per day.

Vitamin C reacts with iodine solution. In this investigation you will use starch as an indicator to determine how much iodine reacts with 50 mg of vitamin C. Then you will use this as a standard to determine the amount of vitamin C in a sample of fruit juice.

Problem

How can you determine the amount of vitamin C in fruit juice?

Materials *(per group)*

Erlenmeyer flask
glass stirring rod
ring stand
clamp
25-mL burette
10-mL graduated cylinder
mortar and pestle
can opener
3 vitamin C tablets, 50-mg
small can of fresh fruit juice
dilute HCl solution
iodine solution
starch solution
distilled water
medicine dropper

Procedure

1. Using the mortar and pestle, crush one 50-mg vitamin C tablet.

2. Wash the powder into an Erlenmeyer flask with 5 mL of distilled water. (Use another 5 mL if necessary to remove all of the vitamin C powder.)

3. Add 2 drops of HCl solution to the flask.

4. Add 2 drops of starch solution.

5. Fill the burette with iodine solution and attach it to the ring stand.

6. Add the iodine solution drop by drop to the flask, swirling the flask each time, until a violet color remains. Record in the Data Table the amount of iodine solution used.

7. Repeat the procedure for two additional trials, record your results, then calculate and record the average value.

8. Use the graduated cylinder to measure 10 mL fruit juice and pour into an Erlenmeyer flask.

9. Repeat steps 3 through 6 for three trials, recording your results in the Data Table. Calculate and record the average value.

Observations

DATA TABLE

Substance	Trial	Iodine used (mL)
50-mg vitamin C tablet	1	
	2	
	3	
	Average	
Fruit juice (10 mL)	1	
	2	
	3	
	Average	

Analysis and Conclusions

1. Using your average values, set up a proportion to determine the amount of vitamin C in your fruit juice sample. The proportion should be

$$\frac{\text{Average value of iodine used for vitamin C tablet (mL)}}{50 \text{ mg}} = \frac{\text{Average value of iodine used for fruit juice (mL)}}{X \text{ mg}}$$

2. What is the amount of vitamin C in your fruit juice in mg per mL? _____
_____

Critical Thinking and Application

1. What is the purpose of determining the amount of iodine that will combine with a 50-mg tablet of vitamin C? _____
_____
_____

2. Why was the volume of iodine used measured at the moment when the contents of the

flask turned violet? _____

_____

_____

3. How might an investigation such as this one have a practical application?

_____

_____

_____

Going Further

Perform a similar investigation to determine the content of vitamin C in frozen fruit juices and in drinks that are only part fruit juice (such as orange drink or lemonade).

Name _____   Class _____ Date _____

_____ *Laboratory Investigation* _____

Observing Blood Circulation

Background Information

Capillaries are the smallest blood vessels in the body. They are made up of a single layer of cells. Capillaries allow materials such as nutrients, oxygen, carbon dioxide, and white blood cells to be exchanged rapidly between body cells and the bloodstream. They also form a bridge between the arteries and the veins.

In this investigation you will examine the circulation of blood in the capillaries of a goldfish's tail.

Problem

How does blood circulate in the capillaries?

Materials *(per group)*

goldfish in an aquarium
5-cm strip of absorbent cotton
petri dish
microscope slide
microscope
medicine dropper

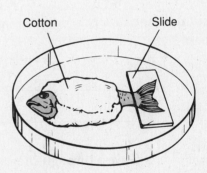

Figure 1

Procedure

1. Saturate the cotton strip with water. Obtain a goldfish from your teacher. Wrap the fish in the wet cotton, leaving the head and tail exposed.

2. Gently place the fish in the bottom half of a petri dish. Spread the tail out. Cover the tail with a microscope slide as shown in Figure 1.

3. Place the dish on the stage of the microscope and examine the tail under low and high powers. **CAUTION:** *Keep the fish moist by adding aquarium water with a medicine dropper to the cotton as it dries out. Remember to use the proper technique for viewing at low and high powers.*

4. Observe the capillaries in the tail. Note the flow of blood through the capillaries. As soon as you have completed your observations and recorded them in Observations, place the goldfish back in the aquarium.

Observations

   1. a. Describe what you see when you focus on the tail membrane under low power.

_____

_____

   b. Does all of the blood flow in the same direction within the same vessel? _____

   Within different vessels? _____

   2. How thick are the walls of capillaries compared to those of arteries and veins?

_____

_____

_____

   3. Do the capillaries vary in size? _____

   4. Does the blood in different capillaries travel at the same speed? _____

Analysis and Conclusions

   1. What is the function of a capillary? _____

_____

_____

_____

   2. What does the flow of blood through the capillaries look like?

_____

_____

_____

   3. Blood travels slowest through the capillaries and fastest through the veins and arteries.

   Why do you think this difference in speed occurs? _____

_____

_____

_____

Critical Thinking and Application

1. Where in the head of the goldfish would you expect to find an extensive network of

   capillaries? Why? _____

   _____

   _____

   _____

2. What do you think happens to a person's blood pressure if something makes the blood
   vessels constrict, or become narrow? Explain your answer.

   _____

   _____

   _____

   _____

3. Eyedrops are sometimes used to remove the redness from the eye. Do you think the
   chemicals in the eyedrops act to narrow or widen the capillaries? Explain your

   answer. _____

   _____

   _____

   _____

   _____

Going Further

Study the effect of temperature on circulation by packing the fish's tail with crushed ice.
Observe the blood flow. After removing the ice, allow the membrane to reach room
temperature, and then add warm water. Record your observations.

_____ *Laboratory Investigation* _____

Circulatory System _____ **9** _____

Investigating the Heart

Background Information

The heart is a fist-sized muscle located to the left of the center of the chest. The heart contains four chambers. The upper chambers are called atria. The lower chambers are called ventricles. Between each chamber, there are valves that prevent the backflow of blood. Blood is carried away from the heart by blood vessels called arteries and carried back toward the heart by blood vessels called veins. Arteries and veins are connected by capillaries. Arteries have muscular, elastic walls to help move the blood through the body. Veins have one-way valves to prevent the backflow of blood on its return to the heart. Oxygen-poor blood from cells of the body enters the heart through the right atrium and is pumped into the right ventricle. The blood then travels into the pulmonary artery, which goes into the lungs. In the lungs, the blood gives off carbon dioxide and picks up oxygen. The oxygen-rich blood returns to the heart by way of the pulmonary vein. The blood enters the left atrium and is pumped into the left ventricle. The blood is pumped out of the heart to cells of the rest of the body through the aorta. The muscular wall of the left ventricle is thicker than the wall of the right ventricle because it has to pump the blood to the entire body.

Each time the ventricles contract, blood is forced through the arteries. This force causes a beat, or pulse, that is felt in arteries at the wrist, neck, and temple. The pulse is exactly the same as the heartbeat.

In this investigation you will examine the chambers and blood vessels of the heart. You will also trace the path of blood through the heart.

Problem

What are the chambers and blood vessels of the heart? What path does blood take through the heart?

Materials *(per group)*

cow's heart (dissected)
dissecting tray
probe
blue pencil
red pencil

Procedure

1. Obtain the dissected cow's heart from your teacher.

2. Rinse the heart with water and place it in a dissecting tray.

3. The right and left sides of the heart are identified according to the side of the animal's body in which each is located. As a result, as you look at it, the heart's right side will be found on your left, and the heart's left side will be found on your right. In Figure 1, label the right and left sides of the heart.

4. Observe the outside of the heart. Locate the blood vessels on the surface of the heart. These blood vessels are the coronary arteries and veins.

5. Locate the two large blood vessels that enter the right atrium. These are the upper and lower vena cavas.

6. Find the blood vessels that leave the right ventricle. These are the pulmonary arteries. Use the probe to trace the four blood vessels that pass behind the heart and empty into the left atrium. These are the pulmonary veins.

7. Locate the valve between the right atrium and right ventricle. Gently squeeze part of the heart and notice how the valve closes.

8. Find the left atrium and left ventricle. Notice the large arched blood vessel that leaves the left ventricle. This is the aorta.

9. In Figure 1, label the right atrium and ventricle, left atrium and ventricle, upper and lower vena cavas, pulmonary arteries, pulmonary veins, aorta, and valves.

10. Color Figure 1 to indicate the flow of blood through the heart. Use the red pencil to show oxygen-rich blood and the blue pencil to show oxygen-poor blood.

11. Use arrows to trace the path of blood through the heart in Figure 1.

Observations

_____ side   _____ side

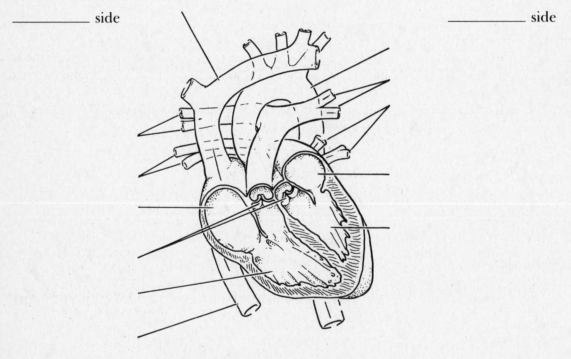

Figure 1

1. Which side of the heart is thicker? _____

2. How many flaps of tissue make up the valves between each of the following:

   a. Right atrium and right ventricle _____

   b. Left atrium and left ventricle _____

Analysis and Conclusions

1. Explain how the coronary blood vessels are important for the functioning of the heart.

   _____

   _____

2. How do the valves control the flow of blood through the heart?

   _____

   _____

3. Which heart chamber has the thickest muscle wall? Explain why.

   _____

   _____

4. Do all arteries carry oxygen-rich blood? Explain.

   _____

   _____

Critical Thinking and Application

1. What would happen if the arteries became narrower?

   _____

   _____

   _____

2. Sometimes babies are born with a hole in the septum, or wall that separates the right and left sides of the heart. Explain how this situation will affect the baby.

   _____

   _____

   _____

   _____

Going Further

   Using reference material, find out what an electrocardiogram is, how it is used, and what information it provides.

_____ *Laboratory Investigation* _____

10 ___

Investigating Breathing and Respiration

Background Information

Breathing is a mechanical process that involves the exchange of oxygen and carbon dioxide. Some animals use lungs for breathing; other animals breathe with gills or through their skins. Respiration is not the same thing as breathing. Respiration is a chemical process. The oxygen that is taken in during breathing is used to burn food in the cells to release energy during respiration. As a result of respiration, the waste products carbon dioxide and water are produced. Carbon dioxide and water are then breathed out.

In this investigation you will compare breathing with respiration. You will also investigate the similarities between the burning of a candle and respiration.

Problem

How much air can your lungs hold? What is the relationship between burning and respiration?

Materials (*per group*)

device for measuring lung
   capacity
glass plate
candle
petri dish
matches
tongs
250-mL beaker
bromthymol blue solution
paper cup
2 straws
oral thermometer
timer
test tube
rubber stopper
test-tube clamp

Procedure
Part A Breathing

1. Working with a partner, record your normal breathing rate for 1 minute. (**Note:** *Breathing in plus breathing out counts as one breath.*) This represents the normal amount of air that moves in and out of your lungs. It is called tidal air.

🔺 **2.** To measure the tidal air that normally moves in and out of your lungs, use the lung capacity device that your teacher has set up for you. To do so, exhale into the tube as shown in Figure 1. Use the markings on the lung capacity device to determine how much water was displaced into the jar. The amount of water displaced, in milliliters, is a measure of your normal lung capacity. Now record your normal lung capacity in Observations.

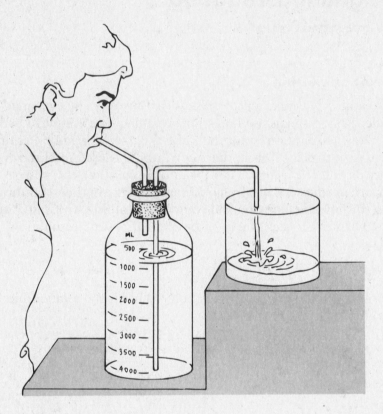

Figure 1

3. Next take a deep breath and force more air into your lungs. This additional air is called complemental air. Exhale as much air as you can. The air that you must force to exhale is known as reserve air.

4. Repeat the measuring of your lung capacity, this time inhaling and exhaling as much air as possible. You have now measured your vital lung capacity. Record this measurement.

5. Your vital lung capacity is one indicator of your physical fitness. To compare your vital lung capacity to a standard capacity for your age and height, do the following:

    a. Compute your height in centimeters.
    b. Multiply your height in centimeters by 20 (for females) or by 23 (for males).
    c. The result is your standard lung capacity in milliliters.

Part B Respiration

    1. Respiration is a chemical process in which food is burned in your cells. This burning is similar to the burning of a candle. Both processes produce heat, water, and carbon dioxide.

🔺   **2.** You can demonstrate that respiration releases heat by taking your body temperature. With a thermometer, take and record your body temperature.

    3. To show that your cells release water as a byproduct of respiration, obtain a cool glass plate. Exhale onto the plate and note the collection of water droplets.

4. To test for the release of carbon dioxide, you will use an indicator called bromthymol blue (BTB). The blue color of the BTB solution will change to green, then yellow, and then colorless when carbon dioxide is bubbled through it.

5. Obtain a cup or beaker about one-third full of water. Add a few drops of BTB. Place a straw in the BTB solution. While your partner times you, gently blow through the straw into the BTB solution. Record the time needed for the solution to become colorless.

6. Obtain a candle, petri dish, and glass plate. Use some melted wax to secure the candle in the center of the petri dish. Light the candle. You can feel the heat that is produced. **CAUTION:** *Be extremely careful when lighting and using the candle.*

7. Using tongs, carefully hold a cool glass plate over the burning candle. Note the water droplets that collect on the surface of the plate. A burning candle releases water.

8. Place the test tube in the clamp and invert it. Hold the inverted test tube completely over the burning candle and keep it there until the candle flame has gone out. Once the smoke from the candle has filled the test tube, use your free hand to quickly place the rubber stopper in the test tube. Turn the test tube right side up.

9. To test for the presence of carbon dioxide, remove the rubber stopper from the test tube, quickly add about 8 drops of BTB, and place the rubber stopper back in the test tube.

10. Shake the test tube gently to mix the BTB and gases. The BTB solution will change to green or greenish-yellow if carbon dioxide is present in the test tube.

Observations
Part A

1. Normal breathing rate = _____ breaths/minute.

2. a. Normal lung capacity (tidal air) = _____.

   b. Vital lung capacity (tidal air + complemental air + reserve air) = _____.

3. How much greater than normal lung capacity is your vital lung capacity?

   _____

4. Can your lungs be completely emptied of air? _____

   _____

5. a. Your height = _____ cm

   b. Your height × 20 or 23 = _____

   c. Your standard lung capacity = _____ mL

6. How does your vital lung capacity compare with the standard you have just computed?

   _____

Part B

1. What is your body temperature in degrees Fahrenheit? _____

   In degrees Celsius? _____

2. What is the time needed for the BTB solution to turn colorless during step 5?

   _____

3. Did the BTB change color during step 10? _____ What does this indicate? _____

   _____

Analysis and Conclusions

1. How is cellular respiration similar to burning a candle?

   _____

   _____

   _____

2. What is the difference in the lung capacity of males as compared to females?

   _____

   _____

   _____

3. Is your lung capacity different from other class members of your own sex? What might

   cause these differences? _____

   _____

   _____

Critical Thinking and Application

1. What kinds of illnesses do you think could cause decreased lung capacity?

   _____

   _____

2. How might lung capacity be increased? _____

   _____

   _____

3. Do you think athletes or nonathletes would have a greater lung capacity? Why?

_____

_____

_____

4. How do you think smoking might affect lung capacity?

_____

_____

5. How would lung capacity be important to some musicians?

_____

_____

Going Further

If a bell jar is available, construct a model of human lungs as shown in Figure 2. To make the model "breathe," pull down and then push up on the rubber sheet. Describe what happens. How does the working of this model relate to the working of your diaphragm and lungs?

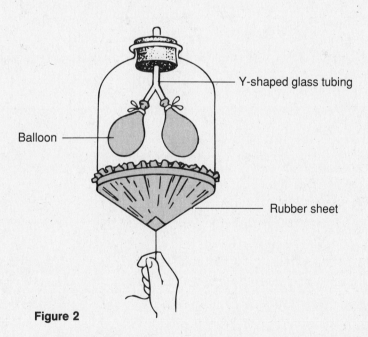

Y-shaped glass tubing

Balloon

Rubber sheet

Figure 2

_____ *Laboratory Investigation* _____

Investigating the Effect of Exercise on Respiration

Background Information

In multicellular organisms (humans, for example), specialized organs and systems are used to exchange the gases involved in respiration with the external environment. The function of the respiratory system is to take oxygen (O_2) into the body and to release carbon dioxide (CO_2) and water (H_2O) from the body. The presence of carbon dioxide can be detected with the testing solution bromthymol blue. A change of color in this solution indicates the presence of carbon dioxide.

In this investigation you will learn how to detect the presence of carbon dioxide with bromthymol blue. You will see that people produce carbon dioxide while breathing. You will also discover the effect that exercise has on the amount of carbon dioxide produced.

Problem

How can the release of carbon dioxide be determined? How is the production of carbon dioxide affected by exercise?

Materials *(per group)*

200 mL of bromthymol blue solution (BTB)	8 drinking straws	medicine dropper
2 500-mL jars or flasks	graduated cylinder	classroom clock or watch with a second hand
glass stirring rod	65 mL of dilute ammonia	

Procedure

Part A   At Rest—Measurement of the Amount of Carbon Dioxide Exhaled

1. Pour 100 mL of bromthymol blue solution (BTB) into a jar or flask.

2. Using a straw, breathe out normally into the BTB solution for exactly 1 minute.
   CAUTION: *Be careful not to suck the solution into your mouth.*

3. The BTB solution should turn a pale yellow at the end of a minute. If it doesn't, continue exhaling until the yellow color appears. Record the amount of time it takes the color change to occur.

4. With a medicine dropper, add a drop of ammonia to the jar and stir once.

5. Continue to add ammonia 1 drop at a time. Count each drop and stir once between drops until the solution turns blue.

6. In the Data Table, record the number of drops of ammonia needed to turn the solution blue.

7. Repeat steps 2 through 6 two times, using the BTB solution. Record your data in the Data Table and find the average number of ammonia drops needed for the three trials.

Part B After Exercise—Measurement of the Amount of Carbon Dioxide Exhaled

1. Pour 100 mL of BTB solution into a second jar or flask.

2. Stand up and exercise for 1 minute.

3. Using a straw, breathe into the BTB solution for exactly 1 minute. Again, be careful not to suck on the straw.

4. When the BTB solution has changed color, add ammonia with a medicine dropper, 1 drop at a time. Stir after each drop. Record the total number of drops needed to turn the solution blue for each of the three trials. Find the average number of drops needed for the three trials.

Observations
Part A

What is the time needed to change solution from blue to yellow? _____

Parts A and B

DATA TABLE

	At Rest	After Exercise
Trial 1		
Trial 2		
Trial 3		
Average		

Analysis and Conclusions

1. In Part A, did it take others the same amount of time to turn the solution from blue to yellow? If not, why might their times be different?

_____

_____

2. What does the number of ammonia drops represent?

_____

3. Are the average numbers of ammonia drops used to restore the blue color in Parts A and

B the same? If not, why are they different? _____

_____

_____

4. What effect does exercise have on the amount of carbon dioxide released from the lungs?

_____

_____

Critical Thinking and Application

1. Why does exercise cause you to breathe faster? _____

_____

_____

2. What are some of the benefits of exercise to your body?

_____

_____

3. Design an experiment to measure the respiration rate of another animal.

_____

_____

_____

_____

_____

Going Further

Find out if germinating seeds give off carbon dioxide during respiration. Obtain about 10 to 15 germinating lima bean seeds. Place them in a flask, or small jar, and add enough bromthymol blue solution (BTB) to cover about half of the seeds. Stopper the flask, or if a jar is used, place a lid on it. Place the jar aside for 30 minutes and then observe the results. What color should the BTB solution turn if carbon dioxide is given off by the seeds?

Laboratory Investigation

12

Observing Human Reflexes

Background Information

An organism is constantly responding to stimuli in its environment. Whether a stimulus is physical or psychological, the nervous system automatically produces a reflex, or response. Some reflexes coordinate internal body processes, such as the slowing or quickening of a heartbeat. Other reflexes protect the organism. For example, the human eye shuts automatically, or flinches, when an object approaches it.

In this investigation you will observe two reflexes.

Problem

What is a reflex?

Materials *(per two students)*

black construction paper,
   12 cm × 17 cm
1 sheet of notebook paper
pencil
watch or clock with sweep
   second hand
small flashlight

Procedure
Part A The Pupillary Reflex

1. Have your partner face a light source, such as a window or a bright light in the classroom.

2. Instruct your partner to close both eyes and to cover them with his or her hands for 45 seconds. **Note:** *Do not press the hands tightly against the closed eyes.*

3. Have your partner uncover and open both eyes while you closely observe the pupils of his or her eyes. Observe any changes in the pupils of the eyes.

4. Have your partner look out the window or at a bright light. Then shade your partner's eyes with a sheet of black construction paper and observe any changes in the pupils. Remove the sheet of paper, and then continue to observe any changes in the pupils.

5. Have your partner hold an edge of the black construction paper vertically to his or her face. The edge of the paper should run along a line from the middle of the forehead, along the bridge of the nose, to the center of the chin. Using the small flashlight, shine a light into your partner's left eye. Observe any changes in both eyes. Then shine a light into your partner's right eye. Observe any changes in both eyes.

6. Switch roles with your partner and repeat steps 1 through 5.

Part B   The Knee-Jerk Reflex

1. Have your partner sit on a chair and cross his or her legs at the knees.

2. Using the side of your hand, tap your partner sharply, but not painfully, just below the kneecap of the top leg, as shown in Figure 1. Observe the response of the tapped leg.

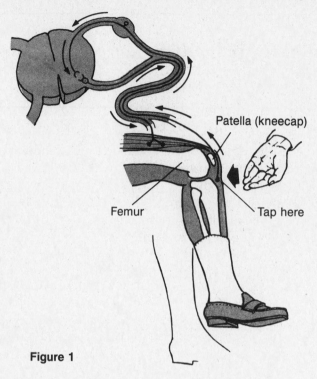

Figure 1

3. On a sheet of paper, write a column of 10 three-digit numbers. Repeat steps 1 and 2 while your partner adds this column of numbers.

4. Repeat steps 1 and 2, while your partner concentrates on not allowing any response to your tap.

5. Switch roles with your partner and repeat steps 1 through 5.

Observations
Part A

1. After your partner opens his or her eyes, what happens to the pupils?

_____

_____

2. a. What happens to the pupils when the sheet of black construction paper is inserted between the light source and your partner's eye?

   _____

   _____

   b. What happens to the pupils when the construction paper no longer blocks your

   partner's eye from the light source? _____

   _____

3. What happens to the pupils when you shine a light into only the left eye? The right eye?

_____

_____

_____

Part B

1. What is the response when you first tap your partner below the knee?

_____

_____

2. What is the response when you tap your partner below the knee while your partner is

adding the numbers? _____

_____

Analysis and Conclusions

1. What are two purposes of a reflex response? _____

_____

_____

2. Explain the changes in pupil size observed in Part A.

_____

_____

3. In Part B, does the response change when your partner tries to control the response?

Explain your answer. _____

_____

_____

Critical Thinking and Application

1. List three other reflex responses in humans.

_____

_____

2. Why do you think a snake is still able to move even after it has been cut into smaller

pieces? _____

_____

_____

3. What effect might alcohol have on a person's reflexes?

_____

_____

Going Further

1. Develop and conduct simple experiments to observe other reflex responses of humans.

2. Design experiments that will test the reflex responses of other animals.

─────── *Laboratory Investigation* ───────

Observing the Effect
of Adrenaline on *Daphnia*

Background Information

Adrenaline is a hormone that is released from the two adrenal glands located on top of the kidneys. Among other things, adrenaline causes the heart to beat faster, increases blood pressure, instructs the liver to release sugar into the blood, and increases the breathing rate. It also causes the activities of the digestive system to slow down.

Daphnia, the water flea, is a small, transparent animal. Many of its body functions, such as heartbeat, can easily be studied under the microscope.

In this investigation you will observe the effect of the hormone adrenaline on a *Daphnia's* heartbeat.

Problem

What effect does adrenaline have on the heartbeat of a *Daphnia?*

Materials *(per group)*

drinking straw
toothpick
petroleum jelly
adrenaline solution
 (about 0.01%)
medicine dropper
microscope
microscope slide with coverslip
metric ruler
Daphnia culture
scissors
clock or watch with second hand

Procedure

 1. Cut off the tip of a drinking straw about 1 mm from the end to form a straw ring.

 2. With a toothpick, put some petroleum jelly along the top rim of the 1-mm straw ring.

 3. Place the ring, jelly side down, on the slide. This ring will act as a chamber for the *Daphnia.*

 4. With a medicine dropper, add a drop of *Daphnia* culture to the chamber. Place a coverslip over the chamber. Observe the slide under the low power of the microscope. Locate one *Daphnia* and notice the location of the heart as shown in Figure 1.

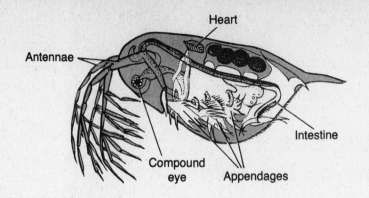

Figure 1

5. Use a watch or clock with a sweep second hand to count the number of heartbeats you observe in 1 minute. Record your data in the Data Table.

6. Remove the slide from the microscope and, with a medicine dropper, add 1 drop of adrenaline to the chamber.

7. Replace the coverslip over the chamber. Observe the slide under the low power of the microscope. Immediately locate one *Daphnia* and count the number of heartbeats per minute. Record this number in the Data Table.

8. Give the *Daphnia* to your teacher for proper disposal. Clean the slide and coverslip.

Observations

DATA TABLE	Beats per Minute
Before adding adrenaline	
After adding adrenaline	

Analysis and Conclusions

1. How does the addition of adrenaline affect the heartbeat of *Daphnia*?

_____

2. Is the effect of adrenaline on *Daphnia* similar to the effect of adrenaline on humans?

Explain your answer. _____

_____

_____

3. What was the reason for the petroleum jelly chamber in this investigation?

_____

_____

Critical Thinking and Application

1. Adrenaline is sometimes called the "fight-or-flight" hormone. Explain why this name is

   appropriate. _____

   _____

   _____

2. In addition to increasing heart rate, breathing rate, and blood sugar, adrenaline also causes the arteries to expand so that more blood can flow to the muscles in your arms and legs. At the same time, adrenaline causes the muscles in your digestive system to slow

   down. How does this help in an emergency? _____

   _____

   _____

3. In what situation might a doctor inject a patient with adrenaline?

   _____

   _____

   _____

Going Further

1. Increase the concentration of the adrenaline solution and study its effect on the heart rate of a *Daphnia*.

2. Devise an experiment in which you can determine the effect of temperature on the heart rate of a *Daphnia*. Upon approval by your teacher, perform the experiment. Describe your results.

Laboratory Investigation

14

Observing the Development of Frog Eggs

Background Information

In the early stages of development of vertebrate animal embryos, the embryos are nearly identical. In later stages, distinct characteristics in the embryos can be observed. The development of embryos can be affected by various outside factors—for example, temperature.

In this investigation you will observe a developing frog egg. You will also try to find out whether cold temperatures affect the development of frog eggs.

Problem

What changes do frog eggs undergo during development? How do variations in temperature affect development and hatching of frog eggs?

Materials *(per group)*

1 or 2 *Elodea* plants	pond water
frog eggs	2 small glass jars
aged tap water	microscope slide
gravel	microscope

Procedure

1. Label one glass jar A: Frog Eggs and the other jar B: Refrigerated Frog Eggs.

2. Fill jars A and B with a mixture of half aged tap water and half pond water. Place some gravel in the bottom of each jar. Anchor the *Elodea* plants in the gravel. Place about 30 frog eggs in each jar.

3. Place jar A in a place where the temperature will remain between 18°C and 20°C. Place jar B in the refrigerator. During the next two weeks, examine these jars every day.

4. Obtain a few frog eggs from your teacher and examine them under the microscope. Make drawings of these eggs in Plate 1. Record the magnification.

5. Remove one or two eggs each day from jars A and B and examine them with a microscope or hand lens. In Data Table 1, record any changes you observe in the eggs.

6. In Data Table 2, record the number of tadpoles that hatch each day over the two-week period.

7. Return all eggs and tadpoles to your teacher.

Observations

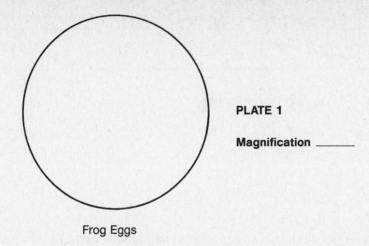

PLATE 1

Magnification _____

Frog Eggs

DATA TABLE 1

	Day 1	Day 2	Day 3	Day 4	Day 5	Day 6	Day 7	Day 8	Day 9	Day 10
Jar A Room Temperature										
Jar B Refrigerated										

DATA TABLE 2

	Day 1	Day 2	Day 3	Day 4	Day 5	Day 6	Day 7	Day 8	Day 9	Day 10
Jar A Room Temperature										
Jar B Refrigerated										

1. What color are the frog eggs? What portion of the eggs is the embryo?

   _____

2. Are all the eggs the same color? Explain. _____

   _____

3. Which jar produced more tadpoles? _____

4. How long did it take the eggs in jar A to turn into tadpoles?

   _____

Analysis and Conclusions

1. What do you think is the function of the jellylike substance that covers the eggs?

_____

_____

2. Describe the process of the early development of a frog embryo.

_____

_____

_____

_____

3. How does temperature affect the hatching of frog eggs?

_____

_____

Critical Thinking and Application

1. Did all the eggs in jar A develop? If not, why? _____

_____

_____

2. What do you think would happen to the frog eggs if they were frozen for a period of

time and then put back into a pond? _____

_____

_____

3. During what season do you think most frog eggs in nature hatch into tadpoles? Explain

your answer. _____

_____

Going Further

Investigate the embryonic development of three animals of different vertebrate classes. Using reference books, prepare a chart that compares the similarities and differences among the embryonic developments of the animals chosen. Use pictures or drawings for each stage of development. Label the various body parts as they appear in each stage. At what point do the embryos begin to show distinct differences in their development?

_____ *Laboratory Investigation* _____

15

A Model for Disease Transmission

Background Information

Infectious diseases are caused by microorganisms such as bacteria and viruses. Most disease-causing microorganisms do not move from one person to another on their own. Instead, the microorganisms are transmitted through contact with an infected person or a contaminated object or substance.

In this investigation you will study how easily harmless microscopic organisms are transmitted from person to person.

Problem

How easily can infectious disease be spread?

Materials *(per four students)*

4 nutrient agar plates
 with covers
4 glass-marking pencils
safety goggles
sterile cotton swabs
sterile water
yeast culture
wire loop
Bunsen burner

Procedure

1. Assign a number from 1 to 4 to each group member.

2. Using a glass-marking pencil, have each group member draw a line on the bottom of an agar plate to divide it in half. Then have each person label one side "Control" and the other side "Experimental." Each person should also mark his or her number on the bottom of the plate.

3. Have the group members swab their left hands with a sterile cotton swab moistened in sterile water. Then have them swab one corner of the control side of the agar with the same swab.

4. Have each group member sterilize a wire loop by passing it through the flame of the Bunsen burner until the entire length of the wire has been heated to a red glow. **CAUTION:** *Put on your safety goggles whenever you use a Bunsen burner.*

5. Have the group members streak their plates with the sterilized wire loops, as shown in Figure 1. The streak should begin from the point at which the plate was touched with the swab. **Note:** *The wire loop should be sterilized after each use by heating as before.*

6. Your teacher will swab the right hand of group member 1 with a culture of yeast.

7. Have group member 1 then shake hands with group member 2. Have group member 2 then shake hands with group member 3. Finally, have group member 3 shake hands with group member 4.

8. Have the group members swab their right hands with sterile cotton swabs moistened with sterile water. Then, using the same swab, have them swab one corner of the experimental side of each agar plate and repeat procedures 4 and 5.

9. Wash your hands thoroughly after swabbing.

10. Have the group members cover all plates and incubate them in an inverted position at room temperature for 48 hours. Then the plates should be examined.

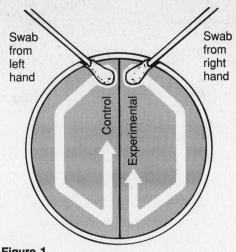

Figure 1

Observations

1. Compare the experimental and control sides of the plates. Describe the difference

   between the two sides. _____

   _____

   _____

2. Why is it necessary to compare your plates with the yeast culture?

   _____

   _____

Analysis and Conclusions

1. How are most disease-causing microorganisms transmitted?

   _____

   _____

2. How were the yeast colonies transmitted from plate 1 to plates 2, 3, and 4?

   _____

   _____

   _____

3. Which plate contains the greatest number of yeast? Which plate contains the least number? Explain your answer.

_____

_____

_____

Critical Thinking and Application

1. What do you think is the best way to prevent the transmission of infectious diseases?

_____

_____

_____

2. How would population density influence the transmission of diseases?

_____

_____

_____

3. List three infectious diseases that are commonly spread among human populations.

_____

_____

_____

Going Further

How can the spread of disease be prevented? Test the sensitivity of the bacteria in your plates to antibiotics. Using forceps, place a pretreated antibiotic disk in the center of each of your petri dishes. Turn the petri dishes upside down and incubate them at 37°C for 24 to 48 hours. When incubation is complete, some areas on the surface of the agar should look cloudy or white. These areas have bacteria growing on them. Check for clear or less dense circular regions around each disk. A clear region around the disk indicates that the antibiotic has either killed bacteria or inhibited their growth.

_____ *Laboratory Investigation* _____

Relating Noninfectious Diseases and Nutrition

Background Information

During your teens, you will grow at a faster rate than at any other time in your life except infancy. This growth involves more than just height and mass. Your bones increase in density, and your muscles develop in size and strength. Your endocrine glands also grow and develop. Good eating habits are especially important during this period of rapid growth.

A well-balanced diet definitely contributes to a healthy body. Yet many teenagers do not have good eating habits. They may skip breakfast, choose snacks that are rich in fats and sugars, go on crash diets, and neglect foods that contain important nutrients. For most Americans, improper nutritional habits cause health problems.

Good eating habits during the teenage years usually mean fewer problems during later years. Scientists have begun to take a closer look at the relationship between nutrient intake and chronic, life-threatening diseases. Heart disease, diabetes, high blood pressure, kidney disease, various digestive disorders, and even certain types of cancer have all been found to be connected with nutrition. It is becoming more and more evident that "you are what you eat."

In this investigation you will plan nutritionally balanced menus.

Problem

Can you plan a menu that meets the Recommended Daily Allowances (RDA) published by the Food and Nutrition Board of the National Academy of Sciences? How closely does your Calorie consumption equal the Calories you burn?

Materials *(per group)*

hand calculator (optional)
pencil
additional tables of nutrition
  information (optional)

Procedure

1. Compute your mass in kilograms (1 lb = 0.454 kg). Compute your height in centimeters (1 in = 2.54 cm). Find your age and height in the Table of Average Masses.

TABLE OF AVERAGE MASSES (kg)

Height (cm)	12 Years M	12 Years F	13 Years M	13 Years F	14 Years M	14 Years F	15 Years M	15 Years F
130	28	29						
134	30	31	30	32				
138	32	33	33	34	33	35		
142	34	35	35	36	35	37	36	
146	37	38	38	38	38	40	38	42
150	40	41	40	41	41	43	41	45
154	43	44	43	45	44	47	45	48
158	46	47	46	48	47	49	47	51
162	48	50	50	51	50	52	51	53
166	52	53	53	54	53	55	55	55
170			55	57	57	58	58	59
174				60	61	60	61	61
178					65	61	65	62
182							68	
186							71	

2. From the foods listed in the Table of Nutritional Information, plan a day's menu. Choose those items you would like to eat for breakfast, lunch, dinner and snacks. List them in the chart entitled My Menu in Observations. Also list the nutritional information for each item.

3. Total each of the columns in your menu chart. Compare the totals with the data in the Recommended Dietary Allowances Table.

RECOMMENDED DIETARY ALLOWANCES

Sex	Age	Calories	Protein (g)	Calcium (mg)	Iron (mg)	Vitamins A (i.u.)	Vitamins B_1 (mg)	Vitamins C (mg)
Males	12–16	2700–3000	46–54	1200	18	5000	1.4	50–60
Females	12–16	2100–2400	44–48	1200	18	4000	1.1	50–60

g = grams, mg = milligrams, i.u. = international units

TABLE OF NUTRITIONAL INFORMATION

Food	Amount	Calories	Protein (g)	Calcium (mg)	Iron (mg)	Vitamins A (i.u.)*	B (mg)	C (mg)
Hamburger	113 g	300	20.2	11	3.1	40	.09	0
Hot dog or hamburger bun	1	90	2.5	22	0.6	trace	.08	trace
T-bone steak	227 g	800	30.0	16	4.4	150	.12	0
Fried chicken	¼	230	22.4	18	1.8	230	.07	0
Egg	11 g	80	6.5	27	1.2	590	.15	0
Flounder	100 g	70	14.9	61	0.8	—	.06	—
Bacon	1 slice	50	1.8	1	0.2	0	.04	0
Hot dog	1	125	7.0	3	0.6	0	.08	0
Shrimp	.2 kg	145	28.3	26	1.8	—	—	—
American cheese	28 g	105	7.0	198	0.3	350	.01	0
Milk	1 cup	160	9.0	288	0.1	350	.07	2
Ice cream	1 cup	255	6.0	194	0.1	590	.05	1
Lima beans	½ cup	130	8.0	28	2.9	—	.12	—
Green beans	½ cup	15	1.0	31	0.4	340	.05	8
Broccoli	1 stalk	25	3.1	88	0.8	2500	.09	90
Corn	1 ear	70	3.0	2	0.5	310	.09	7
Black-eyed peas	½ cup	90	6.5	20	1.7	280	.02	14
Baked potato	1 med.	145	4.0	14	1.1	trace	.15	31
French fries	20	310	4.0	18	1.4	trace	.14	24
Potato chips	20	230	2.0	16	0.8	trace	.08	6
Apple	1 med.	80	0.3	10	0.4	120	.04	6
Banana	1 med.	100	1.0	10	0.8	230	.06	12
Fresh strawberries	½ cup	35	0.5	14	0.6	80	.02	16
Orange juice	1 cup	120	2.0	25	0.2	550	.22	124
White bread	1 slice	65	2.0	22	0.6	trace	.06	trace
Chocolate chip cookie	1	50	0.5	4	0.2	10	.01	trace
Chocolate cake	1 piece	235	3.0	41	0.6	100	.02	trace
Corn flakes	1 cup	95	2.0	6	0.5	0	.10	0
Pancake	1	105	3.2	27	0.6	54	.08	trace
Syrup	1 Tbsp.	50	0	33	0.6	0	—	0

*international units

4. The number of Calories you burn depends, in part, on the activities you perform. Fill in the chart entitled My Activities with the time you spend during an average day on the activities listed.

5. To calculate the total Calories burned in each activity category, multiply A × B × C. Keep the time units the same within each category. For example, if time spent is in hours, use the figure for Calories burned per hour. Add the categories to find the total Calories burned in 24 hours.

Observations

1. a. My weight: _____ kg

   b. My height: _____ cm

   c. My age: _____ years

2.   MY MENU

	Calories	Protein (g)	Calcium (mg)	Iron (mg)	Vitamins A (i.u.)	B$_1$ (mg)	C (mg)
Breakfast							
Lunch							
Dinner							
Snacks							
TOTALS							

3.

MY ACTIVITIES

Activity	A Calories Burned per Hour or Minute	B Minute or Hour Spent in Activity	C Your Mass	Total Calories
Sleeping Napping	.0075/min .45/hr			
Reading Watching TV Eating Sitting in class	.0108/min .64/hr			
Dressing Showering Driving car	.015/min .90/hr			
Light activity Walking Lab work	.0308/min 1.84/hr			
Moderate activity Gym class Bicycling Dancing Easy jogging	.0395/min 2.37/hr			
Heavy work Swimming Tennis Basketball Wrestling Climbing stairs	.0483/min 2.9/hr			

4. Compare your total Calories burned in 24 hours with the total Calories you would consume according to the menu you planned in step 2 of the Procedure.

Calories burned: _____

Calories consumed: _____

Analysis and Conclusions

1. a. How do the totals from your menu chart compare with the Recommended Dietary

   Allowances? _____

   _____

   b. Does your menu provide too many, too few, or the right number of Calories?

   _____

c. In what areas, if any, is your menu deficient? _____

_____

_____

2. Not all of the recommended nutrients have been included in the charts for this activity. Name some other nutrients, including minerals and vitamins, that should be included in

your diet. _____

_____

3. What portion of a gram is 1 mg? _____

4. Review Observation 4 and compare your total Calories burned in 24 hours with the total Calories consumed. If this is your normal pattern, what conclusion can you draw

regarding your Calorie intake? _____

_____

_____

_____

_____

5. What noninfectious diseases have been associated with poor nutrition?

_____

_____

_____

Critical Thinking and Application

1. What two things can you do to safely lose weight?

_____

_____

_____

2. Of the two weight-loss methods you listed in question 1, which do you think is the better

method and why? _____

_____

_____

_____

3. Many companies now advertise breakfast cereals that are low in sugar. Why do you think it is better to eat a breakfast cereal that is lower in sugar?

_____

_____

_____

Going Further

1. Keep a record of all foods you eat every day for one week. Compute your Calorie intake each day. Compute your average daily Calorie intake for the week. How well does your list of foods eaten meet the Recommended Dietary Allowances? Is your daily food intake well balanced in terms of the four basic food groups? What conclusions can you draw regarding your eating habits?

2. Many food products today have nutritional information listed on the package. Look at this information for some of your favorite foods. How well do these foods meet the requirements for Recommended Dietary Allowances?

3. Obtain nutritional information for the major items sold at your favorite fast-food restaurant. Make a chart or booklet of this information and share it with your classmates.

Laboratory Investigation

Alcohol, Tobacco, and Drugs _____ **17** ___

Effects of Tobacco and Alcohol on Seed Germination

Background Information

Ethyl alcohol, or ethanol, an ingredient in alcoholic beverages, is a drug absorbed directly into the bloodstream from the stomach and small intestine. A drug is any substance that causes a change in body function. Alcohol acts as a depressant to the central nervous system. Depressants decrease the action of the central nervous system by reducing the ability of nerves to transmit impulses between the brain and the rest of the body.

Tobacco leaves contain substances that have been shown by scientists to be harmful to people and animals. The tar in tobacco has been linked to cancer. Tars and other substances in tobacco may also lead to heart and respiratory illnesses.

In this investigation you will test the effect of alcohol and tobacco smoke on seeds. A seed that is alive and healthy will germinate under proper conditions of moisture and temperature. But if a harmful substance is added to the seed's environment, it may change the number of seeds that will germinate.

Problem

How do tobacco and alcohol affect seed germination?

Materials _(per group)_

100-mL graduated cylinder	wide rubber tubing
250-mL graduated cylinder	aluminum foil
250-mL Erlenmeyer flask	cotton
3 50-mL beakers	filter paper
glass-marking pencil	20 filterless cigarettes
faucet aspirator	matches
scissors	25 mL ethanol
forceps	150 mustard seeds
two-hole rubber stopper	metric ruler
glass tubing	4 petri dishes and covers
narrow rubber tubing	

Procedure

1. Construct a smoking machine like the one shown in Figure 1.

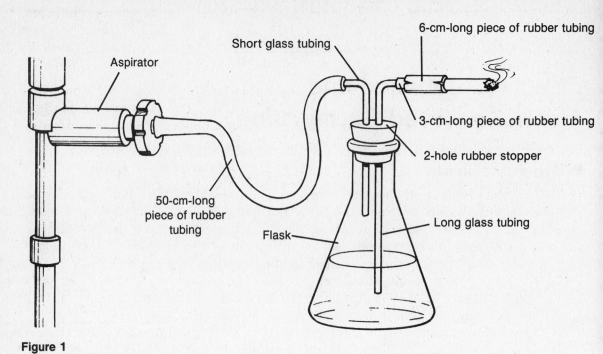

Figure 1

2. Using a graduated cylinder, pour 150 mL of water into a 250-mL flask. Obtain from your teacher a two-hole rubber stopper with glass tubing already inserted. Then place the stopper in the flask. Make sure the longer glass tubing goes into the water. However, the shorter glass tubing should not touch the water.

3. With a gentle, twisting motion, slide the 3-cm-long piece of rubber tubing over the free end of the long glass tubing. Cut a 2-mm notch in each end of the wide, 6-cm-long piece of rubber tubing. To construct your cigarette holder, slide the wide rubber tubing over the narrow rubber tubing.

4. Attach one end of the 50-cm-long piece of narrow rubber tubing to the free end of the short glass tubing. Attach the other end to a faucet aspirator. When the water is turned on, it will produce a partial vacuum that will draw smoke from the cigarette through the water in the flask.

5. Insert a filterless cigarette into the cigarette holder. To catch falling ashes, put a petri dish under the cigarette.

6. Light the cigarette with a match. **CAUTION:** *Be careful when using matches.* Turn on the water. Smoke should bubble through the water. Increase the flow of faucet water through the aspirator if the smoke does not bubble freely through the water.

7. When the cigarette has burned almost to the holder, turn off the water. Using forceps, remove the cigarette, run water over it, and dispose of it in the petri dish with the ashes.

8. Repeat steps 5 to 7 until the smoke from a complete pack of 20 cigarettes has been bubbled through the water.

9. Place 50 mustard seeds in each of three 50-mL beakers.

10. Add 25 mL of water to one of the beakers. Label the beaker. Add 25 mL of ethanol to the second beaker and label it. **CAUTION:** *Keep alcohol away from open flame.* To the last beaker add 25 mL of smoky water and label it. Cover the beakers with aluminum foil and set them aside for 24 hours.

🔥 **11.** Spread a thin layer of cotton across the bottom of each of three petri dishes after 24 hours. Place a piece of filter paper on top of the cotton. Pour the contents of each of the 50-mL beakers into a separate petri dish. Carefully spread the seeds out over the filter paper. Cover the petri dishes and label them. Set the petri dishes aside.

12. After another 24 hours, check the petri dishes for germinated seeds. Count the number of germinated seeds in each petri dish after one day. Record the results in the Data Table. Then set the petri dish aside for another 24 hours. Count the number of germinated seed in each petri dish after two days.

Observations

DATA TABLE

Substance	Number of Seeds Germinated	
	Day 1	Day 2
Water		
Alcohol		
Tobacco smoke		

1. How does the number of germinated seeds in ethanol compare to the number of germinated seeds in tap water? _____

_____

_____

2. How does the number of germinated seeds in smoky water compare to the number of germinated seeds in tap water? _____

_____

_____

Analysis and Conclusions

1. How does alcohol affect the central nervous system?

_____

_____

2. Based on your results, do you think alcohol is harmful to seeds? Explain your answer.

_____

_____

Human Biology and Health

3. To what human illnesses has tobacco been linked?

_____

_____

4. Based on your results, do you think tobacco is harmful to seeds? Explain your answer.

_____

_____

Critical Thinking and Application

1. How might using filtered instead of unfiltered cigarettes change the results of this

experiment? _____

_____

_____

2. If cigarette smoking causes a person's blood vessels to constrict (close up), what will

happen to the person's blood pressure? _____

_____

_____

3. Why should a person who has consumed alcohol not drive an automobile?

_____

_____

_____

4. Caffeine is a substance found in coffee, tea, and chocolate that tends to increase a person's heart rate and breathing rate. Would you consider caffeine to be a drug? Why or

why not? _____

_____

_____

Going Further

1. Call your local chapter of Alcoholics Anonymous or Al-Anon, and request information about the warning signs of alcoholism.

2. Using reference materials, research the medical and social problems created by drug abuse. Choose one drug, give its chemical name, and list its effects on the body. Also find out some of the social problems caused by the selling, buying, and using of this drug.